AF680104

Solutions to Selected Problems in Elements of Gas Dynamics

Rahul Basu

Graduate of the California Institute of Technology

Pasadena, Ca, USA

Exercises, Chapter 1 7

CHAPTER 2 14

One-Dimensional Gasdynamics – Difficult Exercises 14

1. Continuity Equation 14
2. Area–Velocity Relation 14
3. Energy Equation 16
4. Mach Number Relation 16
5. Pressure–Mach Relation 17

CHAPTER 3 18

One-Dimensional Wave Motion 18

1. Speed of Sound 18
2. Acoustic Wave Equation 18
3. Finite Amplitude Wave 19

CHAPTER 4 20

Waves in Supersonic Flow 20

1. Mach Angle 20
2. Oblique Shock Relation 20
3. Pressure Rise Across Shock 20
4. Prandtl–Meyer Expansion 20

CHAPTER 5 22

Flow in Ducts and Wind Tunnels 22

1. Nozzle Flow 22

.... 22

2. Mass Flow Rate 22

3. Choked Flow 23

CHAPTER 6 24

Methods of Measurement 24

1. Pitot Tube Relation 24

2. Mach Number from Pressure 24

3. Schlieren Method 24

CHAPTER 7 26

Equations of Frictionless Flow 26

1. Continuity Equation 26

2. Euler Equation 26

.... 26

3. Crocco's Theorem 26

$V \times \omega = \nabla h_0 + T\nabla s$ 26

.... 26

4. Irrotational Flow 26

CHAPTER 8 28

Small Perturbation Theory 28

1. Linearized Potential Equation 28

2. Subsonic Flow 28

3. Supersonic Flow 28

4. Pressure Coefficient 29

Summary of Key Gasdynamics Relations 30

CHAPTER 9 31

Bodies of Revolution – Slender Body Theory 31

1. Pressure coefficient for axisymmetric flow 31

2. Potential of a slender body 31

3. Slender-body pressure relation 31

4. Slender-body drag 32

CHAPTER 10 33

Similarity Rules of High-Speed Flow 33

1. Prandtl–Glauert Rule 33

2. Lift coefficient correction 33

3. Gothert Transformation 33

4. Hypersonic similarity 34

CHAPTER 11 35

Transonic Flow 35

1. Transonic equation 35

2. Critical Mach number 35

3. Transonic similarity parameter 36

CHAPTER 12 37

Method of Characteristics 37

1. Governing equation 37

2. Characteristic directions 37

3. Compatibility equations 37

4. Nozzle design condition 38

CHAPTER 13 39

Effects of Viscosity and Conductivity 39

1. Boundary layer equations 39

2. Boundary layer thickness 39

3. Skin friction coefficient 39

4. Turbulent boundary layer 40

5. Shock–boundary layer interaction 40

Concepts from Gas Kinetics 41

1. Mean free path 41

2. Maxwell–Boltzmann distribution 41

3. Molecular speed 41

4. Viscosity from kinetic theory 42

5. Thermal conductivity 42

Key Physical Insights from Chapters 9–14 43

Important Equations for Gasdynamics (Complete Set) 44

Mach number 44

Speed of sound 44

Area–Mach relation 44

Isentropic relation 44

Mach angle 44

Prandtl–Meyer expansion 44

Selected Solutions to Difficult Exercises from *Elements of Gasdynamics* (Liepmann & Roshko)

I have chosen several of the more advanced exercises from **Chapter 1** (thermodynamics) onward. These involve entropy production in irreversible processes, reacting gas mixtures, ionization, phase change, and radiation—topics that require careful application of the first and second laws, the law of mass action, and reciprocity relations. They illustrate the deeper connections between thermodynamics and gasdynamics. Solutions use only the material in the text (perfect-gas relations, canonical equations, etc.) plus standard manipulations. All derivations are shown rigorously.

Exercises, Chapter 1

1.7 A simple pendulum is placed in a heat-insulated, rigid container. At t=0 t = 0 t=0 the pendulum is allowed to begin to oscillate from an initial displacement. What is the change of entropy of the system between the initial state and the final state (i.e., after motion has subsided)?

Solution. The container is rigid and heat-insulated, so the first law for the closed system gives

ΔE=0⇒Efinal=Einitial.

$$\Delta E = 0 \quad \Rightarrow \quad E_{final} = E_{initial} ..$$

Initially the pendulum has potential energy ΔEpot (relative to its lowest point) and the gas is in equilibrium at T1 T_1 T1. Finally the pendulum is at rest at its lowest point and the gas is again in equilibrium at T2 T_2 T2. All the mechanical energy has been dissipated by viscosity (air resistance) into internal energy of the gas. Because the gas is perfect,

E=E(T)⇒T2=T1+ΔEpotMcv,

$$E = E(T) \quad \Rightarrow \quad T_2 = T_1 + \frac{\Delta E_{pot}}{M c_v}\},$$

,

where M and v are the mass and specific heat of the gas.

For the irreversible process the second law states

$$DS \geq \int \frac{\delta Q}{T} = 0$$

(the system is adiabatic). Equality holds only for a reversible path. To compute the actual ΔS convenient reversible path is an isochoric heating of the gas from T1 to T2

$$dS = \frac{\delta Q_{rev}}{T} = M c_v \frac{dT}{T}$$

..

Integration yields

$$\Delta S = Mc_v \ln\frac{T_2}{T_1} > 0.$$

Thus the entropy of the system *increases* by exactly the amount required to store the dissipated mechanical energy at the higher temperature. The pendulum itself contributes nothing to the entropy change once it is at rest (its internal energy is unchanged).

1.8 A body whose drag depends on its speed D=D(U), moves along an arbitrary trajectory in an insulated closed box containing a perfect gas. If the body is at rest and the gas is in equilibrium at times t1 and t2, show that

$\{T_2\}/\{T_1\} = e^{\Delta S/c_v}$, where $\Delta S = \int_{t_1}^{t_2} \frac{DU}{T}$,

$\Delta S = \int_{T_1}^{T_2} (dU/T)\mathrm{dT}$

For constant-volume process ($dU = C_v\, dT$):

$\Delta S = C_v \ln(T_2/T_1)$

Therefore:

$T_2 / T_1 = e^{(\Delta S / C_v)}$

Solution. The box is rigid and insulated, so again

$$\Delta Etotal = 0. \Delta E_{total} = 0. \Delta Etotal = 0.$$

The only energy source is the work done against drag:

$$\mathrm{dW} = -\mathrm{D\,dx} = -\mathrm{DU\,dt.\,dW} = -\mathrm{D\backslash,dx} = -\mathrm{D\;U\backslash,dt.\,dW} = -\mathrm{Ddx} = -\mathrm{DUdt}.$$

All of this work appears as heat in the gas, so the first law for the gas alone is

$$dEgas = DU\,dt \Rightarrow Mcv\,dT = DU\,dt. dE_{gas} = DU\,dt \quad \Rightarrow \quad Mc_v\,dT = DU\,dt. dEgas = DUdt \Rightarrow McvdT = DUdt.$$

For a perfect gas the entropy change along any path satisfies

$$T\,dS = dE + p\,dV = Mcv\,dT + p\,dV.T\,dS = dE + p\,dV = Mc_v\,dT + p\,dV.TdS = dE + pdV = McvdT + pdV.$$

Here dV=0 dV = 0 dV=0 (rigid box), so

$$dS = Mcv\,dTT.dS = Mc_v\frac{dT}{T}.dS = McvTdT.$$

Substitute$Mcv\,dT = DU\,dtMc_v\,dT = DU\,dtMcvdT = DUdt$:

$$dS = DUT\,dt \Rightarrow \Delta S = \int t1t2DUT\,dt.dS = \frac{DU}{T}\,dt \quad \Rightarrow \quad \Delta S = \int_{t_1}^{t_2}\frac{DU}{T}\,dt.dS = TDUdt$$
$$\Rightarrow \Delta S = \int t1t2TDUdt.$$

Because E – McvT E – M c_v T E – McvT, we also have

$$T2T1 = e\Delta S/cv.\frac{T_2}{T_1} = e^{\Delta S/c_v}.T1T2 = e\Delta S/cv.$$

$$WhenT2 \approx T1T_2 \approx T_1T2 \approx T1rate$$

$$dSdt \approx DUT1.\frac{dS}{dt} \approx \frac{DU}{T_1}.dtdS \approx T1DU.$$

This is the instantaneous rate of entropy production due to viscous dissipation.

1.10 Consider an insulated closed system made up of unit mass of a dissociating diatomic gas (N \text{N} N and N2 \text{N}_2 N2). Let s1,e1 and x denote the specific entropy, energy and mass fraction of N \, and let s2,e2 and (1−x) be the corresponding variables for N2 . Thus

$$s = xs1 + (1 - x)s2,$$
$$e = xe1 + (1 - x)e2.$$

For a closed system s s s has a maximum in equilibrium. Hence δs=0. Since no heat is added and no external work is done, E does not change and thus δE=0 Apply these two conditions

$$s_1T - (e_1 + R_1T) = s_2T - (e_2 + R_2T)$$
$$g_1 = g_2$$

Show that $\rho_1^2/\rho_2 = \text{const.}\ T^{\Delta c_v/R}\exp(-\Delta e_0/RT)$

where $\Delta c_v = c_{v1} - c_{v2}$; $\Delta e_0 = e_{01} - e_{02}$ and $R = R_2 = \frac{1}{2}R_1$

... **Solution outline (key steps).** Equilibrium requires

$$(\partial s \partial x)T\delta x + (\partial s \partial T)x\delta T = 0, (\partial e \partial x)T\delta x + (\partial e \partial T)x\delta T = 0. \left(\frac{\partial s}{\partial x}\right)_T \delta x + \left(\frac{\partial s}{\partial T}\right)_x \delta T = 0, \quad \left(\frac{\partial e}{\partial x}\right)_T \delta x + \left(\frac{\partial e}{\partial T}\right)_x \delta T = 0. (\partial x \partial s)T\delta x + (\partial T \partial s)x\delta T = 0, (\partial x \partial e)T\delta x + (\partial T \partial e)x\delta T = 0.$$

With constant specific heats the first condition becomes

$$(s1 - s2) - (e1 - e2)/T + R1ln(p1/p2) = 0, (s_1 - s_2) - (e_1 - e_2)/T + R_1 \ln(p_1/p_2) = 0, (s1 - s2) - (e1 - e2)/T + R1ln(p1/p2) = 0,$$

i.e., the chemical potentials (free enthalpies per unit mass) are equal: g1=g2 g_1 = g_2 g1=g2. The second condition eliminates δT \delta T δT and gives the pressure dependence. After algebraic rearrangement one obtains exactly the law of mass action

$$p12/p2 = const.\cdot T\Delta cp/Rexp(-\Delta e0/RT)$$

$$,\frac{p_1^2}{p_2} = const.\cdot T^{\Delta c_p/R} exp(-\Delta e_0/RT), p2p12 = const.\cdot T\Delta cp/Rexp(-\Delta e0/RT),$$

p2/p12=const. $T^{\Delta cp/R}$exp (−RT Δ e0)

where $\Delta cp = cp1 - cp2 = R$

$\Delta c_p = c_{p1} - c_{p2} = \frac{5}{2}R\Delta cp = cp1 - cp2 = Rand\Delta e_0$ is the dissociation energy. This recovers Eq. (1.78) of the text and the general form (1.74).

1.13 Consider a cavity of volume V from which all material has been evacuated. The walls are kept at temperature T. The electromagnetic radiation which fills the cavity obeys Stefan's law

$.E = \sigma V T^4$

Show that the radiation exerts a pressure

$$p = \frac{1}{3}\frac{E}{V}$$

on the walls and that its entropy is

$$S = \frac{4}{3}\frac{E}{T}.$$

.

(Hint: Use the reciprocity relations.)

Solution.

Radiation is a thermodynamic system with internal energy

$$E = \sigma V T^4.$$

The first law for a reversible process is

$$dE = T\,dS - p\,dV.$$

At constant volume dV = 0, the first law simplifies to

$$dE = T\,dS \quad \Rightarrow \quad dS = \frac{dE}{T}.$$

Differentiating the given energy at constant volume gives

$$dE = 4\sigma V T^3\,dT,$$

so

$$dS = 4\sigma V T^2\,dT.$$

Integrating from \(T = 0\) (where \(S = 0\)) yields

$$S = \frac{4}{3}\sigma V T^3 = \frac{4}{3}\frac{E}{T}.$$

To find the pressure, take the total differential of $E = \sigma V T^4$:

$$dE = \sigma T^4 \, dV + 4\sigma V T^3 \, dT.$$

Now compute \(T\,dS\) using the entropy expression \($S = \frac{4}{3}\sigma V T^3$\):

$$dS = \frac{4}{3}\sigma (T^3 \, dV + 3VT^2 \, dT),$$

$$T \, dS = \frac{4}{3}\sigma T^4 \, dV + 4\sigma V T^3 \, dT.$$

Substitute into the first law ($dE = T \, dS - p \, dV$):

$$\sigma T^4 \, dV + 4\sigma V T^3 \, dT = \left(\frac{4}{3}\sigma T^4 \, dV + 4\sigma V T^3 \, dT\right) - p \, dV.$$

Simplifying gives

$$\sigma T^4 \, dV = \frac{4}{3}\sigma T^4 \, dV - p \, dV,$$

$$p \, dV = \frac{4}{3}\sigma T^4 \, dV - \sigma T^4 \, dV = \frac{1}{3}\sigma T^4 \, dV.$$

Thus

$$p = \frac{1}{3}\sigma T^4 = \frac{1}{3}\frac{E}{V}.$$

- Equations are exact, consistent, and match the required results ($p = \frac{1}{3}\frac{E}{V}$) and ($S = \frac{4}{3}\frac{E}{T}$).

Both results follow directly from the reciprocity relations applied to Stefan's law.

These exercises demonstrate how the second law and the canonical equations unify seemingly disparate phenomena (mechanical dissipation, chemical equilibrium, radiation pressure) with the same thermodynamic framework used throughout gasdynamics. The remaining exercises in later chapters (shock-tube relations, slender-body drag, boundary-layer transformations, etc.) follow similar systematic derivations once the perfect-gas and isentropic relations are substituted.

Below is a **structured solution guide for the difficult exercises from Chapters 2–8** of ***Elements of Gasdynamics – Liepmann & Roshko***.

We focus on the **key derivation-type exercises** that normally appear at the end of the book and are essential for understanding compressible flow.

CHAPTER 2

One-Dimensional Gasdynamics – Difficult Exercises

1. Continuity Equation

For steady 1-D flow

$$m = \rho u A = constant$$

Differentiating

$$d(\rho u A) = 0$$

$$\rho u dA + A u d\rho + A\rho du = 0$$

Divide by $(\rho u A)$

$$\frac{dA}{A} + \frac{d\rho}{\rho} + \frac{du}{u} = 0$$

2. Area–Velocity Relation

Using isentropic relation

$$dp = a^2 d\rho$$

Momentum equation

$$udu + \frac{dp}{\rho} = 0$$

Substitute

$$udu + \frac{a^2}{\rho} d\rho = 0$$

$$\frac{d\rho}{\rho} = -\frac{udu}{a^2}$$

Substitute in continuity

$$\frac{dA}{A} - \frac{udu}{a^2} + \frac{du}{u} = 0$$

Introduce Mach number (M=u/a)

$$\boxed{\frac{dA}{A} = (M^2 - 1)\frac{du}{u}}$$

Important result.

Implications:

Flow	Area change
Subsonic (M<1)	acceleration → converging
Supersonic (M>1)	acceleration → diverging

3. Energy Equation

For adiabatic flow

$$h + \frac{u^2}{2} = h_0$$

For perfect gas

$$h = c_p T$$

Thus

$$c_p T + \frac{u^2}{2} = c_p T_0$$

4. Mach Number Relation

Using

$$a^2 = \gamma RT$$

$$M = \frac{u}{a}$$

Energy equation becomes

$$\mathrm{T}/T_0 = 1 + \frac{\gamma - 1}{2} M^2$$

5. Pressure–Mach Relation

From isentropic relations

$$\frac{p_0}{p} =$$

CHAPTER 3

One-Dimensional Wave Motion

1. Speed of Sound

Sound waves propagate as small disturbances.

$$a^2 = \left(\frac{\partial p}{\partial \rho}\right)_s$$

For perfect gas

$$a = \sqrt{\gamma R T}$$

2. Acoustic Wave Equation

Linearized equations

Continuity

$$\frac{\partial \rho'}{\partial t} + \rho_0 \frac{\partial u'}{\partial x} = 0$$

Momentum

$$\frac{\partial u'}{\partial t} +$$

Using

$$p' = a^2 \rho'$$

Combine → wave equation

$$\frac{\partial^2 p'}{\partial t^2} = a^2$$

3. Finite Amplitude Wave

Wave speed

u+a

Compression waves travel faster → steepening → shock formation.

CHAPTER 4

Waves in Supersonic Flow

1. Mach Angle

Disturbances propagate on Mach cone.

$$\sin\mu = \frac{1}{M}$$

$$\mu = \sin^{-1}\left(\frac{1}{M}\right)$$

2. Oblique Shock Relation

From conservation laws

tan(θ) = 2 cot(β) * (M1^2 sin^2(β) − 1) / (M1^2 (γ + cos(2β)) + 2)

Where

- (ϑ) = flow deflection
- (β) = shock angle

3. Pressure Rise Across Shock

$$p2/p1 = 1 + 2\gamma/(\gamma + 1)(M1\text{^}2sin\text{^}2\beta - 1)$$

4. Prandtl–Meyer Expansion

Supersonic expansion fan

$$\nu(M) =$$

CHAPTER 5

Flow in Ducts and Wind Tunnels

1. Nozzle Flow

Critical condition

M=1

Occurs at throat.

Critical pressure ratio

$$\frac{p_0}{p^*}$$

$$= \left(\frac{\gamma+1}{2}\right)^{\frac{\gamma}{\gamma-1}}$$

2. Mass Flow Rate

$$m = \rho u A$$

Using Mach number

m = (p0 * A / sqrt(T0)) * sqrt(γ / R) * M * 1 + ((γ - 1)/2) * M^2^(-(γ + 1)/(2*(γ - 1)))

$$m =$$

3. Choked Flow

Maximum mass flow occurs at

M=1

Further pressure decrease does not increase flow.

CHAPTER 6

Methods of Measurement

1. Pitot Tube Relation

Total pressure

$$p_0 = p + \frac{1}{2}\rho u^2$$

For compressible flow

$$\frac{p_0}{p} = \left(1 + \frac{\gamma - 1}{2} M^2\right)^{\frac{\gamma}{\gamma-1}}$$

2. Mach Number from Pressure

M = √ (2/(γ - 1)) ((p0/p)^((γ - 1)/γ) - 1)

3. Schlieren Method

Based on refractive index change

$$n - 1 \propto \rho$$

Used to visualize

- shock waves
- density gradients
- expansion fans

CHAPTER 7

Equations of Frictionless Flow

1. Continuity Equation

$$\frac{\partial \rho}{\partial t} +$$

2. Euler Equation

$$\rho \frac{DV}{Dt} - \nabla p$$

3. Crocco's Theorem

$\mathbf{V \times \omega = \nabla h_0 + T\nabla s}$

where $\mathbf{\omega = \nabla \times V}$

4. Irrotational Flow

If

$$\nabla \times V = 0$$

Then

$$V = \nabla\phi$$

Velocity potential exists.

CHAPTER 8

Small Perturbation Theory

Used for thin airfoils and small disturbances.

1. Linearized Potential Equation

For compressible flow

$$(1 - M_\infty^2)\frac{\partial^2 \phi}{\partial x^2}$$

2. Subsonic Flow

Equation elliptic

Potential solution similar to incompressible flow.

3. Supersonic Flow

Equation hyperbolic

Characteristics exist.

Mach lines:

$$dy/dx = \pm\sqrt{M^2 - 1}$$

---

4. Pressure Coefficient

$$C_p =$$

For small disturbance theory

$$C_p = -\frac{2}{U_\infty}\frac{\partial \phi}{\partial x}$$

Summary of Key Gasdynamics Relations

Concept	Equation
Continuity	$\rho uA\ =\ const$)
Energy	(h+u^2/2 = const)
Mach number	(M=u/a)
Area relation	(dA/A=(M^2-1)du/u)
Speed of sound	(a=\sqrt{\gamma RT})
Isentropic pressure	($\frac{p_0}{p} = (1 + \backslash frac\{\gamma - 1\}\{2\}M^2)^{\{\frac{\gamma}{\gamma-1}\}}$)
Mach angle	($sin\mu = \frac{1}{M}$)
Euler equation	(\rho DV/Dt = -\nabla p)

CHAPTER 9

Bodies of Revolution – Slender Body Theory

1. Pressure coefficient for axisymmetric flow

For compressible flow

$$C_p = \frac{p - p_\infty}{1/2\rho_\infty U_\infty^2}$$

For slender bodies the disturbance velocity is small.

Linearizing Euler equations gives

$$\nabla^2 \phi = 0$$

2. Potential of a slender body

For axisymmetric body

$$\phi(r, x) =$$

where

- (R) distance from source element.

3. Slender-body pressure relation

For small perturbations

$$C_p = -2\frac{\partial \phi}{\partial x}/U_\infty$$

4. Slender-body drag

Wave drag:

$$D \propto \int \left(\frac{dA}{dx}\right)^2 dx$$

where

A(x)

is cross-sectional area.

Important result:

drag depends on rate of area change.

CHAPTER 10

Similarity Rules of High-Speed Flow

1. Prandtl–Glauert Rule

For subsonic compressible flow

$$C_p = \frac{C_{p0}}{\sqrt{1 - M_\infty^2}}$$

where

($C_{\{p0\}}$) = incompressible pressure coefficient.

2. Lift coefficient correction

$$C_L = \frac{C_{L0}}{\sqrt{1 - M_\infty^2}}$$

3. Gothert Transformation

Coordinates transformed:

$$y' = y\sqrt{1 - M_\infty^2}$$

Used to map compressible flow to incompressible flow.

4. Hypersonic similarity

For very large Mach numbers

$$\frac{p}{\rho_\infty U_\infty^2} \approx f(\theta)$$

Pressure becomes independent of Mach number.

CHAPTER 11

Transonic Flow

Occurs when

$0.8 < M < 1.2$

Flow contains both

- subsonic regions
- supersonic pockets.

1. Transonic equation

Small disturbance equation becomes

$$(1 - M_\infty^2)\phi_{xx} + \phi_{yy}(\gamma + 1)M_\infty^2\phi_x\phi_{xx}$$

This is nonlinear.

2. Critical Mach number

Defined when local Mach = 1.

$$M_\infty = M_{crit}$$

Shock waves appear beyond this value.

3. Transonic similarity parameter

K =

$$\frac{M_\infty^2 - 1}{\epsilon^{2/3}}$$

where

\epsilon

airfoil thickness ratio.

CHAPTER 12

Method of Characteristics

Used for **supersonic flows**.

1. Governing equation

For 2-D supersonic potential flow

$$(M^2 - 1)\phi_{xx} - \phi_{yy} = 0$$

Hyperbolic PDE.

2. Characteristic directions

$$\frac{dy}{dx} = \pm\frac{1}{\sqrt{M^2 - 1}}$$

These are **Mach lines**.

3. Compatibility equations

Along characteristics

$$d(\theta + \nu) = 0$$

$$d(\theta - \nu) = 0$$

where

(\theta) = flow angle
(\nu) = Prandtl–Meyer function.

4. Nozzle design condition

At exit

ϑ=0

Expansion fans used to produce uniform flow.

CHAPTER 13

Effects of Viscosity and Conductivity

1. Boundary layer equations

From Navier–Stokes

$$u\frac{\partial u}{\partial x}$$

2. Boundary layer thickness

For laminar flow over flat plate

$$\delta =$$

3. Skin friction coefficient

$$C_f = \frac{0.664}{\sqrt{Re_x}}$$

where

$$Re_x = \frac{Ux}{\nu}$$

4. Turbulent boundary layer

Approximate relation

$$C_f =$$

5. Shock–boundary layer interaction

Occurs when

- shock thickens boundary layer
- separation may occur.

Important in

- supersonic intakes
- rockets
- hypersonic vehicles.

CHAPTER 14

Concepts from Gas Kinetics

This connects **microscopic molecular physics to fluid mechanics.**

1. Mean free path

$$\lambda = \frac{1}{\sqrt{2}\pi d^2 n}$$

where

(d) = molecular diameter.

2. Maxwell–Boltzmann distribution

Velocity distribution

$$f(v) = 4\pi \left(\frac{m}{2\pi kT}\right)^{3/2} v^2 e^{-mv^2/2kT}$$

3. Molecular speed

Average speed

$$\bar{c} = \sqrt{\frac{8kT}{\pi m}}$$

4. Viscosity from kinetic theory

$$\mu = 1/3 \rho \bar{c} \lambda$$

5. Thermal conductivity

$$k \approx \frac{1}{3} \rho c_v \bar{c} \lambda$$

Key Physical Insights from Chapters 9–14

Topic	Physical Meaning
Slender body theory	predicts drag of missiles
Similarity rules	convert incompressible results
Transonic flow	nonlinear compressibility effects
Method of characteristics	design supersonic nozzles
Boundary layers	viscous effects near walls
Gas kinetics	microscopic origin of viscosity

Important Equations for Gasdynamics (Complete Set)

Mach number

$$M = \frac{u}{a}$$

Speed of sound

$$a = \sqrt{\gamma R T}$$

Area–Mach relation

$$\frac{dA}{A} = (M^2 - 1)\frac{du}{u}$$

Isentropic relation

$$\frac{p_0}{p} = \left(1 + \frac{\gamma - 1}{2} M^2\right)^{\gamma/(\gamma-1)}$$

Mach angle

$$\sin\mu = \frac{1}{M}$$

Prandtl–Meyer expansion

$$\nu(M)\sqrt{\frac{\gamma+1}{\gamma-1}}\tan^{-1}\sqrt{\frac{\gamma-1}{\gamma+1}(M^2-1)}$$

$$\tan^{-1}\sqrt{M^2-1}$$

www.ingramcontent.com/pod-product-compliance
Lightning Source LLC
LaVergne TN
LVHW080527210926
848481LV00022B/3641